BONSAI
A HANDBOOK

THE AUTHOR

Dr. R.K.Roy is a renowned horticulturist in India with specialization in Floriculture and Landscaping. He has worked in reputed organizations/research institutes like The Agri-Horticultural Society of India, Kolkata; CSIR-National Botanical Research Institute, Lucknow in various capacities. Dr. Roy has been deeply involved in R&D work on Floriculture and Landscaping since last 40 years. He has developed 16 new varieties of ornamentals (Bougainvillea, Canna, Gladiolus *etc.*) which are popular in floricultural trade. Moreover, he had developed core collection of many ornamentals *viz.* Bougainvillea, Canna, Cycads, Cacti and Succulents, Gladiolus *etc.* during his service tenure. Considering his outstanding contribution, he was awarded with many national and international awards/fellowships like 'Dr. B.P. Pal Award', 'Commonwealth Fellowship', 'Coke Trust Award', 'RHS Financial Award' by Royal Horticultural Society, London.

Dr. Roy has authored many books *viz.* 'Ornamental Annuals', 'House Plants', 'Kitchen Garden', 'Gladiolus', 'Bougainvillea', 'Marigold', 'Ornamental Trees of India', 'Fundamentals of Garden Designing', ' Bulbous and Rhizomatous Plants', 'Home Garden', 'Ornamental Climbers' *etc.*

BONSAI
A HANDBOOK

by

R. K. Roy

2025

Daya Publishing House®
A Division of
Astral International Pvt. Ltd.
New Delhi – 110 002

ISBN: 978-93-5919-343-4

Published by : **Daya Publishing House®**
A Division of
Astral International Pvt. Ltd.
– ISO 9001:2015 Certified Company –
4736/23, Ground Floor, Ansari Road, Darya Ganj
New Delhi-110 002
Ph. 011-43549197, 23278134
e-mail: info@astralint.com
Website: www.astralint.com

Printed at : **Replika Press Pvt Ltd**

FOREWORD

Bonsai is the art and science of dwarfing plants. It is a technique of growing plants (trees, shrubs and other plants) in shallow containers/ trays with the help of pruning, pinching and training in order to produce a three dimensional effect. It is a kind of living sculpture and idealized nature in miniature.

Bonsai is now popular all over the world and very useful for the people who are living in the flats and have little scope to have outdoor gardens. Every house owner likes to have few specimens for interior decoration. As these occupy very small space, can easily be accommodated in house interiors. Considering its present popularity, Bonsai is now important horticulturally both from commercial and aesthetic point of view.

Commercial importance of Bonsai in horticultural and nursery trade is also enormous. The most common plant of Bonsai is *Ficus* species. The cost of a nursery plant of *Ficus* or any other plant for Bonsai is INR 100.00. However, the same plant after value addition and trained as Bonsai can fetch price much more than the nursery plant. Therefore, the technique has full potentiality for exploitation in more aggressive way. Establishing a bonsai nursery is a successful entrepreneurship in peri-urban areas and a scope for employment generation and self-reliance of the rural youth.

The present book "**Bonsai – A Handbook**" is a comprehensive book on Bonsai culture. It contains 16 chapters on various aspects viz. History,

Principles, Classification, Styles, Containers, Pedestal / Stands, Selection of Plants, Planting, Pruning, Training and maintenance besides Tools & Implements. Commercial importance of Bonsai in nursery trade has also been described. The main purpose of the book is to provide technical information and guidance to make bonsai in successful way by all class of interested persons.

Dr.R.K.Roy, Retd. Sr. Principal Scientist, CSIR-National Botanical Research Institute, Lucknow, who was Head, Botanic Garden, Floriculture & Landscaping Division, has written this book. Dr. Roy has vast experience on the subject and has conducted many training courses in Bonsai during his tenure in CSIR-NBRI and otherwise.

I am sure this book will be purposeful for the students, amateur gardeners, nurserymen, commercial growers and, above all, bonsai lovers.

R.K. Tomar

PREFACE

'Bonsai', the art and science of making plants dwarf, is a popular technique all over the world. Most of the garden lovers like to grow Bonsai in a very fascinating way. Because of its, unusual shape, dwarfness and beauty, these are liked by all. Moreover, display of Bonsai requires comparatively less space than other plants and the impact is much more by exhibiting a striking appearance. Therefore, considering its beauty and utility, Bonsai specimens are ideal for display in indoor as well as outdoor places. To reduce today's environmental problems and improving indoor air quality, Bonsai trees / plants are ideal.

In India, popularity of Bonsai technique and availability of plants in nursery are enormous. Every individual having love for plants like to grow Bonsai. By that way, large number of people likes to know its techniques and grow these in their individual capacity. Majority of such people are amateurs and like to start the hobby afresh, having not much knowledge and skill. This bottleneck often lands them in trouble and frustration. Then they look for training from experts to learn the techniques. Experts often need a manual to explain the technique in a better way.

During organization of hundreds of such training programmes on Bonsai technique by the Author, it was felt that a handbook furnishing the technique with diagrams could be useful for the beginners as well as others, interested to the subject. Non-availability of such a handbook, necessitated taking up writing this book which will fulfill the need for the Bonsai lovers. The book contains specific technical and scientific information on various aspects starting from history, technique details regarding planting, maintenance as well as display etc., spread over 16 chapters. Commercial importance of Bonsai in nursery trade has also been described, as an entrepreneurship. The main purpose of the book is providing technical information and guidance to make bonsai in a successful way by all class of interested persons.

The information provided in this book is based on scientific inputs derived from various sources of national and international journals, books and web sources. This has also been supplemented with field studies, growing in Conservatories and, finally, the observations made on live plants besides practical experience of the author over 40 years, in national and international conditions.

I am sure this book will be useful for all the people looking for bonsai technique, as a part of their hobby or otherwise. I sincerely thank the management of Astral International Pvt. Ltd., New Delhi for accepting my proposal for publication of this book for the benefit of garden lovers.

I am grateful to Dr. R.K.Tomar, Director, Directorate of Horticulture & Food Processing, *Uttar Pradesh* for writing the "FOREWORD" of this book. I also like to thank one and all, who directly or indirectly supported me, for writing this book.

R.K. Roy

CONTENTS

1

INTRODUCTION

Bonsai is the art and science of dwarfing plants and growing them in shallow containers / trays. The word 'Bonsai' is a Japanese term consisting of two words, '*Bon*' – a shallow container / tray and '*sai*' – a plant. Therefore, bonsai is a technique of growing plants (trees, shrubs and other plants) in shallow containers/trays with the help of pruning, pinching and training in order to produce a three dimensional effect. It is a kind of living sculpture and idealized nature in miniature. An ideal bonsai is a replica of particular plant or an artistic variation in shape whose different parts are in complete harmony. Development of bonsai requires an artistic eye, skills on training, pruning and knowledge on growing plants in containers. It is the result of a set of cultural techniques (pruning, pinching, clipping, wiring and training) and its practices which help the plants to grow in limited space in an artistic way. Preparation of Bonsai needs an artistic eye and technical know-how for the creation. The beauty of the Bonsai is adorned by all section of people specially those living in cities, where there is no space for growing greeneries. Interiors *viz.* bed room, central hall, reception, window-sill, gallery *etc.* are ideal spaces for display.

Bonsai specimens are dwarf, potted trees that are part of the Japanese culture and very well-known in the world. The technique is a living art and represents peace and tranquility besides long-lasting determination and patience. Originally from China, Bonsai technique dates back hundreds of years and are well known for their miniature and cute appearance. It was Japanese who first presented bonsai to the world during World Trade Fair in 1878 in Paris and subsequently in London in 1909.

This technique was developed by the Chinese over 2000 years ago and later excelled by the Japanese. The technique was started during Heian dynastic period (794-1191 A.D.) and subsequently popularized in 1800 A.D. Bonsai is now popular all over the world and very useful for the people who are living in the flats and have little scope to have outdoor gardens. Every house owner likes to have few specimens for interior decoration. Considering its present popularity, Bonsai is now important horticulturally both from commercial and aesthetic point of view.

Bonsai are of different types and can be grown in a variety of ways. They require daily grooming, but not require high maintenance. Unlike what most people believe, keeping a Bonsai tree alive and thriving isn't difficult. They require simple attention and need to be attended at regular intervals.

Modern Bonsai specimens are combination of various things – plants, containers, support rocks/stones, base decoration materials which give a complete look of a natural landscape in miniature form. Bonsai specimen should be very attractive, distinctive in its style, form and decoration. One should keep in mind that Bonsai is part horticulture and part art. Collectively, Bonsai is a living sculpture in which many components are put together in a harmonious way in order to create an attractive art work with a live plant.

One has to be very selective about the plants keeping in mind the place of display – indoor or out door. Accordingly, styles, support, decoration *etc.* has to be finally decided. Display of well developed bonsai specimen is another important task to draw full attention to the objects. Above all, right technique, style, and combination of many things create a perfect Bonsai after hard work and patience of many years.

2

HISTORY

The Bonsai technique was developed by the Chinese over 2000 years ago. It was developed as a part of pleasure garden display of the Royal Tang dynasty during 618-907 A.D. The technique was further improved with the passage of time. The specimens thus prepared were displayed in interior decoration of the royal family. In China, Bonsai were of two forms – 'Penjai' (refers a tree planted in a container / tray without base landscape decoration) and 'Penjing' (denotes a tree planted in a container / tray with base decoration by landscape).

Later, in Japan the Bonsai technique excelled during Heian period. Bonsai is now popular all over the world and very suitable to the people living in the flats of multi-storied building, as they do not have outdoor gardens. Every house owner likes to use Bonsai specimens for interior decoration. Considering its present popularity, Bonsai is now important horticulturally both from commercial and aesthetic point of view. Brief history of development and popularization all over the world is given below.

Africa

South Africa – Activities on Bonsai in the form of show and establishment of Bonsai Club was started during early sixties. Gradually Bonsai became popular and people liked the culture and became curious. By 1980, international experts started visiting the clubs and promoted activities related to Bonsai. Local nurseries and amateur growers started growing and commercial activities started for sale purpose by using plants locally available.

Zimbabwe – First Bonsai society was founded in1969 marking the formal beginning of Bonsai culture. Evening meetings and training programmes besides other related activities were started to popularize the Bonsai Culture.

Asia

Bonsai culture was much popular in Asian countries than other continents, particularly in the following countries.

Indonesia – The popularity of Bonsai was much in Indonesia and near by islands. First Indonesian Bonsai Society was started in 1979 which was a centre of activities on Bonsai. Locally available plant material was first preference for Bonsai making. Thereafter, new plants species was chosen and the technique was much popular with the passage of time.

Philippines – Bonsai in Philippines is also poplar now. The activities started way back in 1972 by a few bonsai hobbyists and establishment of Philippine Bonsai Society in 1972. Meetings and exhibitions were regularly organized in different parts of the country to popularize the Bonsai culture.

India – The interest on ornamental plants in various forms in India was started with the establishment of Royal Indian Botanic Garden in 1787. Subsequently, creation of The Royal Agri-horticultural Society of India (AHSI) in 1820 by William Carey started Bonsai culture in India. Mr. S. Percy Lancaster, the then Secretary of AHSI made valuable contribution for popularization of Bonsai. Bonsai culture was introduced in India during second part of the twentieth century. Primarily centered at Delhi, the culture gradually spread to Karnataka, Maharashtra, West Bengal. First Bonsai club was established by Mr. S. Srinivas, named as 'Bonsai Srinivas'. In many Indian cities Bonsai culture started and popularized during 1960-70. A remarkable Bonsai garden was established in Lalbagh Botanical Garden, Bangalore, Karnataka in 2008 where more than 2,000 specimens were displayed. The contribution of Mr. V.P. Agnihotri, Delhi who developed some very attractive Bonsai, was significant and widely appreciated. The details of the technique like containers, potting mixture and pruning were introduced and taught by Mrs. Mary Case, USA which lead to the formation of Indian Bonsai Society in 1972 by Mrs. S. Vaswani, Bombay. Further, the formation of Bonsai Study Group of the Indo-Japanese Association in 1979 co-founded by Nikunj and Jyoti Parekh created a great awareness on Bonsai technique and growing plants in dwarf forms in various ways. Their efforts in the form of study tour, workshop, invited lecture, demonstration by experts through out India popularized Bonsai in a magnificent way. Subsequently, this study group organized several workshops, training programmes, exhibitions which created several hundreds serious bonsai growers all over India. Books on Bonsai were also published by Indian authors to provide technical information on the technique. National Bonsai Park was established at Lodhi Garden, New Delhi in 1996.

China – The love for Chinese people for nature and its manifestation in various forms created an interest for Bonsai making which dated back to ancient time. Further, popularization and use of Bonsai were made during Yin and Zhou Dynasty besides Jin Dynasty (220-420 A.D.). Further improvement and maturity came during Tan Dynasty around 907 A.D. Flourishing activities of Chinese *Penjing* work appeared in the Ming and Qing Dynasties (1368-1911 A.D.). Chinese have a fascination of decorating places, hotels, public places/gardens with *Penjing*. As result, *Penjing* gardens are seen all over the country particularly Guang Zhou Xi Yuan Garden, Shanghai Botanical Garden and in other places.

Japan – The history of Bonsai in Japan is very old. The modern Bonsai technique and it's popularization around the world is the contribution of Japan. During 6th century, Japanese Imperial and Buddhist students visited China and brought containers, plants and other related materials essential for Bonsai preparation. Zen Buddhism promoted Bonsai culture through various teachings. Growing trees in Bonsai form and further decoration imitating nature started around 14th century. The technique so well cherished and adopted that the oldest living bonsai tree (more than 500 years old) considered a 'National Treasure' of Japan placed in Tokyo Imperial Palace Collection.

People interested to look the carefully crafted creations, a work for many years' passionate efforts, may take a tour to 'Omiya Bonsai Village',

Saitama, first publicly run collections. There are currently, five gardens running since 1800. Many bonsai exhibitions/festivals are organized every year, where masterly crafted specimens are displayed.

Australia / New Zealand

First Bonsai Society in Australia was formed in Sydney in 1965. After few years in 1971, Bonsai Society of Victoria was formed which was followed in other states. Establishment of Bonsai Federation of Australia was a landmark for increasing popularity. In New Zealand, first Society was formed in Auckland in 1967 which paved the way for introduction and popularization of Bonsai in this country. Bonsai technique is now very popular in Australia/New Zealand.

Europe

Bonsai was first displayed in Europe at the great exhibition in Paris in 1878 and subsequently in London in 1909. But public awareness about Bonsai in Europe was created when large and attractive specimens were displayed in Chelsea Flower Show. Gradually, the culture spread to The Netherland, Germany, Switzerland, France and United Kingdom through club activities. 'European Bonsai Association was established at Heidelberg in 1980 and the bonsai culture entered into many other European countries.

United Kingdom – The awareness regarding Bonsai was created during 1909 when an attractive specimen from Japan was displayed in the Flower Show in London. Subsequently, some plant specimens and trays were imported from Japan by affluent people who were interested towards Bonsai. First sensation was created when a perfect specimen of Bonsai was displayed in Chelsea Flower Show in 1962. Later on 'Federation of British Bonsai Society' was established having objective to popularize Bonsai culture amongst the common people. With this objective, first convention was organized in 1983. Gradually, the Bonsai techniques became popular and demand of the plants was increased together with training on the technique. Now, it is very popular in Britain.

Russia

In 1976, Russian Academy of Sciences Main Botanical Garden got some miniature trees through Japanese Embassy which created lot of interest about dwarf plants and their artistic look. With the establishment of a Japanese Garden in 1987 in Moscow, people got acquainted with miniature trees and their use in gardening. Subsequently, many TV shows, exhibitions *etc.* were organized to popularize Bonsai culture. Ultimately, in 1989 Bonsai Club was started followed by exhibitions, shows which created interest among the common people about Bonsai.

3

PRINCIPLES

The ultimate purpose of growing Bonsai is to create a miniaturized but realistic representation of nature in the form of a dwarf tree. Plants for Bonsai are not genetically dwarf. In fact, any tree species can be used to grow as Bonsai. The best Bonsai – whether a single tree or a multi-plant or rock landscape composition show us something new and instigate us to touch, take notice, stop us as they catch our experience and imaginations. Thick trunks, textured bark, an interplay of twisting live wood and deadwood, surface roots, fine branch and twig ramification, foliage pads, relatively small leaves or needles, a complementary and relatively shallow container, tiny fruit or cones or flowers – these are just a few of the more obvious features that can be used to help portray a miniature landscape. A true Bonsai master knows what to combine and his or her creation attract us. Thus a masterpiece created pleases us when first looked at them and momentarily take away our breath.

While planning for Bonsai, one should have in mind the future image of the grown up specimen on maturity. Moreover, it is to take into account that Bonsai components are part live and growing while others are static and compliment in nature. It is also to be considered that long life and natural changes to be taken place gradually. Therefore, Bonsai specimens can, in fact, live longer than their full-size counterparts because of increased attention to their health, water and nutritional needs, protection from weather extremes, injuries, or pest infestations requiring treatments.

Techniques such as pinching of buds, pruning and wiring of branches, and carefully restricting nutrients ultimately maintain the health of the plant. Most commonly an ideal specimen is kept under a meter in height. However, plants with smaller leaves do make these compositions easier to design. In fact, any plant species that has a woody stem or trunk, grows true branches is ideal. Moreover, if it can be successfully grown in a container to restrict its roots / food storage capability, and has smaller or reducible-leaves, well adopted to create a Bonsai.

For a starting material, you may look around at trees, bushes, hedges in the yard or park, plants in the nursery or wild landscape. These need to be collected carefully during the appropriate growing or dormant season. Most native plants can then be grown in the local climate, as indoor or outdoor.

4

CLASSIFICATION

There is wide variation in shape, size, ornamentation, look, height, spread, and plant species of Bonsai. For academic and practical purpose, it is useful to categorize the specimens for use. Primarily, Bonsai is classified depending upon the height into following four groups.

- **Miniature** (less than 15 cm. in height) - Bonsai in this category is also known as "Mame". Containers are also very small but the specimen looks vary attractive and highly priced. It is a special creation by combining art and horticultural practices. Special care is to be taken for maintenance.
- **Small** (up to 30 cm) – Little bigger in size than "*mame*" and morphological expressions like trunk, branches, leaves *etc.* are more visible.
- **Medium** (30-60 cm) – In Japan it is also known as "*Midi*", bigger than earlier two categories.
- **Large** (60-100 cm) - It is relatively easy to handle and maintenance is not very difficult. Life span is more than other categories.

There may be variation again under the Miniature group according to the plant species, shape and decoration. Similarly, in small category one can find variability with respect to shape, size and ornamentation. Medium and small categories are largely available and much popular than the other two categories.

The ultimate goal of Bonsai is to create a realistic depiction of nature. As it gets smaller (even down to a few centimeters), Bonsai increasingly becomes abstract, as opposed to resembling nature in a more precise way. Several classifications of Bonsai have been put forward, and although the exact size classifications are disputed, they help to gain understanding of the aesthetic and botanical aspects of Bonsai.

Common Names and Bonsai Size/Classes (As per Japanese System)

Large Bonsai

Common Name	Size/Class	Height
'Imperial Bonsai'	Eight-handed	1.5–2.0 m
'Hachi-uye'	Six-handed	1.0–1.5 m
'Dai'	Four-handed	0.75–1.2 m
'Omono'	Four-handed	0.75–1.2 m

Medium-size Bonsai

Common Name	Size/Class	Height
'Chiu'	Two-handed	0.40–0.90 m
'Chumono'	Two-handed	0.40–0.90 m
'Katade-mochi'	One-handed	0.25–0.45 m

Miniature Bonsai

Common Name	Size/Class	Height
'Komono'	One-handed	15–25 cm
'Shohin'	One-handed	13–20 cm
'Mame'	Palm size	05–15 cm
'Shito'	Fingertip size	05–10 cm
'Keshitsubo'	Poppy-seed size	03–8 cm

Purpose

The main purpose of classification is to broadly select the bonsai specimen for various usages. Miniature and small bonsai specimens are generally used for table decoration and gift purpose being small in size and shape. These have special attraction due their curious shape and size. Generally, these are planted in small trays having no scope for base decoration.

Bonsai in medium category are very popular for use as these create significant visual impact by way of base decoration, plant species, style and pedestal. In nursery trade these fetch good price and very useful for decoration in house interiors.

On the other hand, large Bonsai specimens are in great demand in corporate world. These are relatively costly than the other two groups but they create unique aesthetic effect when properly displayed.

Therefore, categorizations of bonsai specimens are useful considering their shape, size and visual look. Commercialization is also dependent on classified specimens.

5

STYLES

There are several styles to give Bonsai a definite shape. The artistic view of the specimens is actually depended on the style followed. Therefore, it is necessary to know the different styles and the technique involved for the development. There are more than 30 different styles on the basis of framework. However, 5-7 styles are popular and exclusively used. Some popular styles have been described so that one can understand and follow the techniques for bonsai development. In Japanese Bonsai style, there are around 11 different styles.

The popular styles which are commonly followed are given below.

- **Straight Trunk** – Single trunk, trained in upright direction, no curve is allowed.
- **Curved Trunk** – Single trunk but curved, giving a curious look.
- **Formal Upright** – This style of training is characterized by a straight, upright, tapering trunk. Branches progress regularly from the thickest and broadest at the bottom to the finest and shortest at the top. Stem (single), straight with lateral alternate branches on both sides, gradually tapering to the tip.
- **Informal Upright** – Stem (single), not straight, incorporating visible curves in trunk and branches but informally curved with lateral alternate branches on both sides. The apex of the informal upright is located directly above the trunk's entry into the soil line

Formal Upright

Informal Upright

- **Slanting Trunk** – The main trunk slant either left or right from the container at an angle. Arrangement of branches is little different with more number of branches on the top.
- **Semi-Cascade** – Trunk horizontal, branches grow informal way leaning towards left or right. Containers with depth are required for this purpose.
- **Cascade** – A unique style, main stem grows straight up to certain height and abruptly trained downwards in cascade form. Containers with depth are required for this purpose.

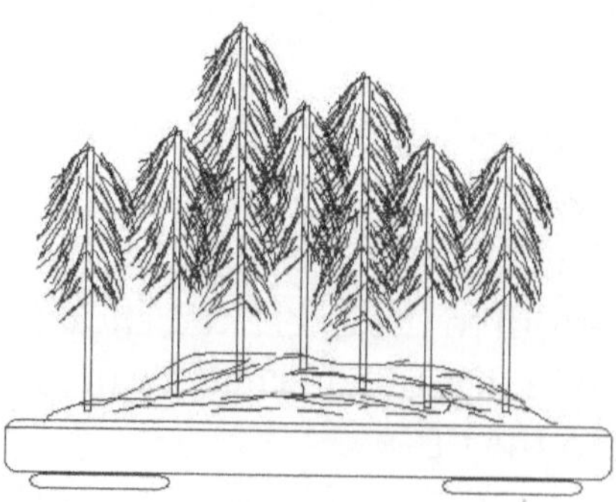

Group Planting

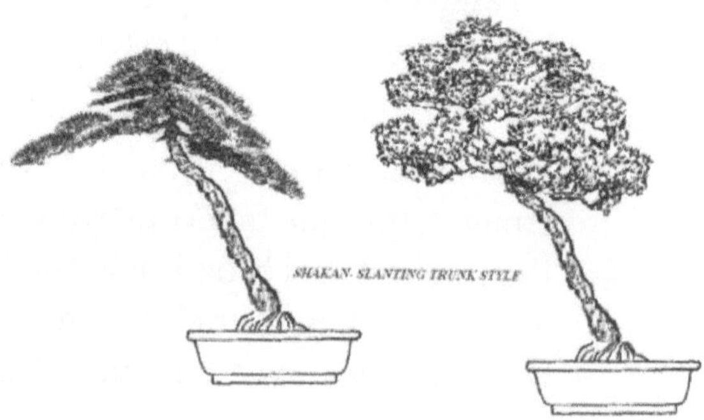

Slanting Trunk

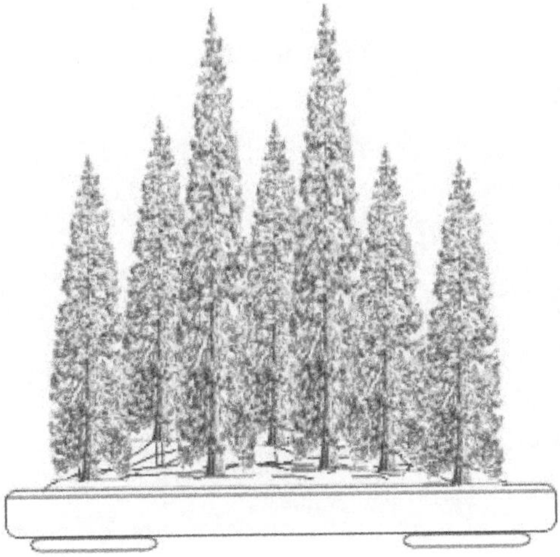

Group Planting (Forest Look)

Wind Swept Style

- **Multiple Trunks** – In this style multiple trunks are allowed to grow by planting more than one plant in the tray. The number of trunks varies depending upon the size of the container.
- **Group Planting** – A style consisting of many plants in a single tray, may be same species or different, aesthetically arranged in combination with stones, gravels *etc.* to make the arrangement intriguing.
- **Broom/Umbrella Style** – Single trunk but multiple branches, trained upward, giving a look of a broom.
- **Raft Style** – Container, plant and style all combined in a special fashion so that it looks like a raft.
- **Wind Swept Style** – A style copied from the nature. Branches of the plant are trained in one direction, seen during storm, little difficult style but attractive.
- **Rock Hugging** – Plant and rocks in combination in a unique form, harmoniously blended together in proportionate way.
- **Root-over-rock** - A style in which the roots of the tree are wrapped around a rock, entering the soil at the base of the rock.
- **Growing-in-a-rock** - A style in which the roots of the tree are growing in soil contained within the cracks and holes of the rock.

Rock Hugging

Multiple Trunk

- **Forest** or **Group** - A style comprising the planting of several or many trees of one species, typically an odd number, in a bonsai pot.
- **Literati** - A style characterized by a generally bare trunk line, with branches reduced to a minimum, and foliage placed toward the top of a long, often contorted trunk.

6

TRAYS/CONTAINERS

Bonsai refers a plant planted in shallow containers. Therefore, containers must be shallow in depth. Various kinds of containers are used for growing bonsai. Clay, porcelain, cement, wooden is common and has different looks and advantages. Unconventional containers made of rocks, wood or any other material suitable for planting is also used considering its beauty and utility. Shape of the containers are also varies like round, oval, rectangular, square *etc.* The size also varies from small to large depending upon the specimen plant. The containers must have drainage holes and toes for better drainage and aeration. Colour of the containers should not be much glazy instead in natural colour and hues *viz.*, grey, brown *etc.*

The final selection of the trays/containers is generally made considering the plant species, style, growth habit and ultimate size of the specimen. Cascade style is better displayed in a container having considerable depth. Square, rectangular shape is ideal for upright style while group plantation is better accommodated in oval trays/containers.

There are alternatives to the conventional ceramic pot. Multi-tree bonsai may be created atop, a fairly flat slab of rock, with the soil mounded above the rock surface and the trees planted within the raised soil. In recent times, bonsai creators have also begun to fabricate rock-like slabs from raw materials including concrete and glass-reinforced plastic. Such constructed surfaces can be made much lighter than solid rock, can include

Various Types of Pots/Containers

depressions or pockets for additional soil, and can be designed for drainage of water, all characteristics difficult to achieve with solid rock slabs. Other unconventional containers can also be used, but in formal bonsai display and competitions in Japan, the ceramic bonsai tray is the most common container.

For bonsai being shown formally in their completed state, container shape, color and size are chosen to complement the tree as a picture frame is chosen to complement a painting. In general, containers with straight sides and sharp corners are used for formally shaped plants, while oval or round containers are used for plants with informal

Various Shapes of the Trays/Containers

designs. Many aesthetic guidelines affect the selection of pot finish and color. For example, evergreen bonsai are often placed in unglazed pots, while deciduous trees usually appear in glazed pots. Trays are also distinguished by their size. The overall design of the bonsai tree, the thickness of its trunk, and its height are considered when determining the size of a suitable container.

7

PEDESTALS/STANDS

The beauty of a perfect Bonsai is expressed when it is displayed in proper way. Depending upon the size, style and place of display, pedestal/stands are required to keep the plant specimen. These may be different in height, shape, size depending upon the specimen to be displayed.

The bonsai display stands are used for the aesthetic purpose of raising the bonsai pot above the surface or floor. This gives a sense of dignity and prominence as well as being highly decorative. Bonsai and *'Penjing'* look best when displayed on a stand or bonsai turntable (bonsai display tables that rotate 360 degrees to optimize view-ability).

Before selection of pedestals/stands following points should be considered.

1. Place of display – Outdoor or indoor.
2. Type of display – Single or in group.
3. Height of display – According to arrangement pattern and size of the specimens.
4. Selection of pedestals / stands - Wooden, Metal, Stone, Cement concrete.
5. Size and Shape – Depending upon the specimen and number of plants to be displayed. Shapes may be round, rectangular, bench type *etc.*

6. Display Pattern – Informal (curved line) or Formal (straight line).
7. Visibility – Should be placed in such a pattern that all specimens are visible to visitors from different angle.

The proper technique for arranging bonsai for public viewing is almost an art itself. It requires enough experience and knowledge to learn the full technique of display. The term 'proper display' is relative, because the environment in which the bonsai will be displayed must be considered. In a home, there is one way to display Bonsai. Other display system is followed when displayed in deck or in the garden. If the Bonsai is to be displayed with others in a club show at the local shopping mall, the arrangement will be different. If it will be seen alone in a special viewing area such as an alcove or a *tokonoma*, then shall be treated entirely in a different way. Each of such viewing locations must be taken into account and the arrangement and inclusion (or exclusion) of various display elements adapted to the particular situation.

Correct placement of elements in a viewing alcove is perhaps the best way to discuss the principles necessary for proper presentation. It allows all various design elements in one isolated location so that they can be clearly understood.

The Western version of a '*tokonoma*' is also a box, but the size and dimensions can vary great.

The elements use in construction of a "*tokonoma*" or alcove display is usually limited to three following principles.

❀ **The Focal Element** – Here the focal element is the bonsai placed on any stand, table or *diaza* on which it is mounted. It must be remembered that it is the primary object is display. Therefore, care must be taken so that main focus is not diverted.

❀ **Companion Element** – This is referred to as the accent piece or accent plant accompanying the focal element – bonsai. This should be well harmonized and complement to each other. It may be another smaller bonsai, an accent plant, an arrangement of mosses *etc.*

❀ **The Distant View Element** –It is usually a two dimensional object for providing background for the composition. Traditionally, with the Japanese, it has been a scroll painting. With a Western approach,

items other than scrolls may be used. These might include a painting or drawing with or without a contemporary frame. It must be remembered that the item used should compliment and harmonize with the focal element of the composition.

A *tokonoma* display is almost always thematic in some way, although but not necessarily. Themes might reflect the seasons, or any special holiday such as Christmas or Thanksgiving *etc.*

Whatever theme is chosen, the elements used in final composition should reflect following three aspects.

❀ **Relevance** – The objects selected for each element should relate to each other in a logical and natural context. A bonsai displaying springtime flowers might be displayed with a photograph of a bird or a scroll depicting the branch of a blooming tree. A deciduous tree displayed without leaves in the winter time might be shown with a scroll painting. A bonsai in fall foliage color might be accented with a small basin stone. The combinations possible artistic are limitless, but they should be selected with deference to entire composition.

- **Sub-ordinance** – When the focal point is a bonsai, there should be any other items near by which creates sub-ordinance and distract attention from the main object. Both near and distant view elements along with such other variables as pot selection and choice of display stands, need to defer and be subordinate to the tree itself.
- **Balance** – Relevance, sub-ordinance and balance to be combined in such a way that they have an asymmetrical balance. Apply the same kind of criteria for arranging elements in a display. Triangulate the composition and avoid placement of any elements so that there is no obstruction of view of the other elements. Generally elements should not be placed in the center of the display space. Bonsai, the focal element, is commonly placed midway into the space and not to the either side of the specimen.

8

SELECTION OF PLANTS

Making Bonsai is a specialized task. Therefore, one has to be very selective in each step from beginning, as every step influence the final outcome. Selection of right type of plant is extremely important. Following points should be considered before selection.

- Purpose – Commercial or normal display.
- Condition - Indoor or outdoor placement.
- Type – Flowering or foliage.
- Size – Small, medium, tall.
- Style – Suitability of the selected plant.
- Climatic Zones – Temperate, sub-tropical or tropical.

Once above points are considered, you may look for sources of availability and collection.

Source and Collection

Plants suitable for making Bonsai are easily available in commercial nursery. Selection of right genus, species and variety as per requirement is a basic thing. To display the characteristic of a true bonsai – age, style *etc.*, the plant selected should be fit enough with the purpose. It should be evaluated that after a reasonable time, the selected plant specimen will grow into a desired thing.

- **Propagated Plants** - If the desired plant species is not available, propagation (layering/cuttings) may be done from the existing mother plant to have a true to type species.
- **Nursery Grown** - Direct purchase from nursery stock or from any reliable sources/gardens/plant sale centres.
- **Other Sources** - Sometimes, look for commercial bonsai grower who sale pre-mature and semi-finished plant which could easy to train.
- **Collection from Wild** - Collection can also be done from natural source *viz.* wild, arboretum, natural habitat. Collection is also made from naturally grown plants in the crevices, especially *Ficus* species.
- **Grown from Seeds** - Plants raised from seeds are preferred.

Selection

A wide range of plant species is selected for bonsai. Selection of plants, its use and a particular style for training depend on the choice person concerned. The ultimate purpose is to create an attractive specimen in dwarf form. Following plant species are suitable for Bonsai purpose.

Trees

Sl.No.	Botanical Name	Family	Common Name
1.	*Abis pindrow*	Pinaceae	'Himalayan Fir'
2.	*Acacia angulta*	Fabaceae	'Black Wattle'
3.	*A. nilotica*	Fabaceae	'Gum Arabic Tree'
4.	*A. cinerea*	Fabaceae	'Chinese Lantern', ' Bell Mimosa'
5.	*A. procera*	Fabaceae	'White Siris, Karoi Tree'
6.	*A. nilotica*	Fabaceae	'Egyptian Acacia'
7.	*Acer buergerinaum*	Sapindaceae	'Trindent Maple'
8.	*Achras sapota*	Sapotaceae	'*Sapota*', ' Noseberry'
9.	*Adenanthera pavonina*	Mimosaceae	'Barbados Pride', 'Sandalwood Tree'
10.	*Adenium obesum*	Apocynaceae	'Desert Rose', 'Impala Lily'
11.	*Adansonia digitata*	Malavaceae	'Baobab Tree', 'Monkey Bread Tree'
12.	*Albizzia lebbeck*	Fabaceae	'Lebbek Tree'

Sl.No.	Botanical Name	Family	Common Name
13.	*Anona squamosa*	Annonaceae	'Sugar Apple', 'Custard Apple'
14.	*Averrhoa carambola*	Oxalidaceae	Carambola Tree', 'Star Fruit'
15.	*Azadirachta indica*	Mileaceae	'Neem Tree', 'Margosa Tree', 'Indian Lilac'
16.	*Barringtonia acutangulaa*	Lecythidaceae	'Indian Oak'
17.	*B. racemosa*	Lecythidaceae	'Powder-puff Mangrove', 'Barringtonia'
18.	*Bougainvillea* spp./cvs	Nyctaginaceae	'Paper Flower'
19.	*Bambusa ventricosa*	Poaceae	'Buddha's Belly Bamboo'
20.	*Bauhinia hookeri*	Fabaceae	'Pegunny', 'Hooker's Bauhinia'
21.	*Brownea ariza*	Fabaceae	'Handkerchief Plant'
22.	*B. coccinea*	Fabaceae	'Mountain Rose, ' Scarlet Flame Bean'
23.	*Brassia actiniphylla*	Araliaceae	'Australian Umbrella Tree'
24.	*Butea monosperma*	Fabaceae	'*Palash*', 'Flame of the Forest'
25.	*Bursera delphichiana*	Burseraceae	'Indian Lavender', 'Bursera'
26.	*Buxus microphylla* 'Kingsville Dwarf'	Buxaceae	'Kingsville Dwarf'
27.	*Caesalpinia cacalaco*	Fabaceae	'Mexican Bushbird', 'Cascalote'
28.	*C. coriaria*	Fabaceae	'Dvi-Dvi'
29.	*Carpinus laxiflora*	Betulaceae	'Aka Shide'
30.	*Cedrus atlantica*	Pinaceae	'Atlas Cedar'
31.	*Celastrus punctatus*	Celastraceae	'Staff Vine', 'Bitter Sweet'
32.	*Callistemon lanceolatus*	Myrtaceae	'Bottle Brush'
33.	*C. polandii*	Myrtaceae	'Gold-tipped Bottle Brush'
34.	*Carpinus turczaninovii*	Betulaceae	'Korean Horn Bean'
35.	*Casuarina equisetifolia*	Casuarinaceae	'Costal Sheoak', ' Horsetail She-Oak'
36.	*Carissa carandus*	Apocynaceae	'Karonda', ' Bengal Currant'
37.	*Celtis africana*	Cannabaceae	'White Stinkwood'
38.	*C. sinensis*	Cannabaceae	'Chinese Hackberry'
39.	*Chamaecyparis lawsoniana*	Pinaceae	'Port Oxford Cedar'
40.	*Citrus x aurantifolia*	Rutaceae	'Key Lime', 'West Indian Lime'
41.	*Citrus microcarpa*	Rutaceae	'*Nagengi*', 'Philippine Lime'

Sl.No.	Botanical Name	Family	Common Name
42.	*Cowania mexicana*	Rosaceae	'Mexican Cliffrose'
43.	*Crataegus oxycantha*	Rosaceae	'European Hawthorn'
44.	*Cryptomeria japonica*	Cupressaceae	'Japanese Cidar', ' Sugi'
45.	*Cupressus sempervirens*	Cupressaceae	'Mediterranean Cypress'
46.	*C. funebris*	Cupressaceae	'Chinese Weeping Cypress'
47.	*Dalbergia sisso*	Fabaceae	'Indian Rosewood', 'Shisham'
48.	*Diospyros kaki*	Ebenaceae	'Japanese Persimon'
49.	*Eleagnus pungens*	Elaegnaceae	'Silver Thorn', 'Thorny Olive'
50.	*Erythrina crista-galli*	Fabaceae	'Cockspur Coral Tree'
51.	*E. indica*	Fabaceae	'Indian Coral Tree'
52.	*Eucalyptus citriodora*	Myrtaceae	'Lemon Scented Eucalyptus'
53.	*Eugenia jambolana*	Myrtaceae	'Malabar Plum, 'Java Plum'
54.	*Ficus benghalensis*	Moraceae	'Banyan', '*Bargad*'
55.	*F. benjamina*	Moraceae	'Weeping Fig'
56.	*F. infectoria*	Moraceae	'White Fig', '*Philkhan*'
57.	*F. microcarpa*	Moraceae	'Indian Laurel, 'Curtain Fig'
58.	*F. tsila*	Moraceae	'Indian Bat Fig;
59.	*F. religiosa*	Moraceae	'Sacred Fig, '*Peepal*'
60.	*F. rubiginosa*	Moraceae	'Rusty Fig', 'Port Jackson Fig'
61.	*F. rumphi*	Moraceae	'Golden Rumph's Fig', 'Golden Mock Bodhi Tree'
62.	*F. triangularis*	Moraceae	'Triangle Fig'
63.	*F. retusa*	Moraceae	'Retuse Fig'
64.	*F.* 'Long Island'	Moraceae	'Dwarf Weeping Fig'
65.	*Fortunella japonica*	Rutaceae	'Kumquat'
66.	Ginkgo biloba	Ginkgoaceae	'Gingko',
67.	*Gliricida maculata*	Fabaceae	'Mexician Lilac'
68.	*Grevillea robusta*	Proteaceae	'Southern Silky Oak', 'Silky Oak'
69.	*Houttynia cordata*	Saururaceae	'Chameleon Plant'
70.	*Jacaranda mimosaefolia*	Bignoniaceae	'Blue Jacaranda'
71.	*Juniperus chinensis*	Cupressaceae	'Chinese Juniper'
72.	*J. prostrata*	Cupressaceae	'Juniper'
73.	*J. rigida*	Cupressaceae	'Juniper'
74.	*J. virginiana*	Cupressaceae	'Juniper'

Sl.No.	Botanical Name	Family	Common Name
75.	*Kigelia pinnata*	Bignoniaceae	'African Sausage Tree'
76.	*Lagerstroemia flosreginae*	Lythraceae	'Pride of India'
77.	*Larix laricina*	Pinaceae	'American Larch', 'Eastern Larch'
78.	*Liquidamber styraciflua*	Altingaceae	'American Sweetgum', 'Hazel Pine'
79.	*Mangefera indica*	Anarcadiaceae	'Mango'
80.	*Melia azadarach*	Meliaceae	'Persian Lilac', 'Indian Lilac'
81.	*Pemphis acidula*	Lythraceae	'Bantigue'
82.	*Picea pungens*	Pinaceae	'Blue Spruce', 'Green Spruce'
83.	*P. jezoensis*	Pinaceae	'Ezo Spruce', 'Dark-bark Spruce'
84.	*Parmentia cerifera*	Bignoniaceae	'Candle Tree'
85.	*Parthnocissus tricuspidata*	Vitaceaae	'Boston Ivy'
86.	*Pinus densiflora*	Pinaceae	'Japanese Red Pine', 'Korean Red pine'
87.	*P. longifolia*	Pinaceae	'Chir Pine', 'Himalayan Pine'
88.	*P. thunbergii*	Pinaceae	'Japanese Back Pine'
89.	*Poinciana regia*	Fabaceae	'Gulmohor'
90.	*Polyalthia longifolia*	Annonaceae	'Ashoka'
91.	*P. aungusta*	Annonaceae	'Ashoka' (Small Leaved)
92.	*Pongamia pinnata*	Fabaceae	'Indian Beech', 'Karanj'
93.	*Pseudocydonia sinensis*	Rosaceae	'Chinese Quince'
94.	*Pithacolobium saman*	Fabaceae	'Rain Tree'
95.	*Prosopis cineraria*	Mimosaceae	'Khejri Tree'
96.	*Prunus mume*	Rosaceae	'Chinese Plum', 'Japanese Plum'
97.	*Putranjiva roxburghii*	Putranjivaceae	'Putranjiva', 'Lucky Bean Tree'
98.	*Psidim guajava*	Myrtaceae	'Guava'
99.	*Quercus infectoria*	Fagaceae	'Alpee Oak'
100.	*Rhododendron arboricum*	Ericaceae	'Tree Rhododendron'
101.	*Saraca indica*	Fabaceae	'Asoka Tree', Ashok'
102.	*Salix babylonica*	Salicaceae	'Babylon Willow', 'Weeping Willow'

Sl.No.	Botanical Name	Family	Common Name
103.	*Streblus asper*	Moraceae	'Toothbrush Tree', 'Siamese Rough Bush'
104.	*Styrax japonica*	Styracaceae	'Sycamoreleaf Snowball'
105.	*Sequoiadendron giganteum*	Cupressaceae	'General Sherman Tree', 'Giant Redwood'
106.	*Tamarindus indica*	Fabaceae	'Tamarind'
107.	*Tamarix parviflora*	Tamaricaceae	'Small Flower Tamarindus'
108.	*T. junipernia*	Tamaricaceae	Small Flower Tamarindus'
110.	*Taxus baccata*	Taxaceae	'English Yew', 'European Yew'
111.	*T. suga*	Taxaceae	'English Yew', 'European Yew'
112.	*T. heterophylla*	Taxaceae	'English Yew', 'European Yew'
113.	*T. martensianaa*	Taxaceae	'English Yew', 'European Yew'
114.	*T. occidentalis*	Taxaceae	'English Yew', 'European Yew'
115.	*Wrightia religiosa,*	Apocynacaea	'Sweet Indrajao'
116	*Ulmus chinensis*	Ulmaceae	'Chinese Elm'
117	*Zelkova serrata*	Ulmaceae	'Japanese Elm', 'Japanese Zelkova'
118	*Zizuphus jujube*	Rhamnaceae	'Chinese Date', 'Red Date'

Shrubs

Sl.No.	Botanical Name	Family	Common Name
01.	*Acocanthera spectabilis*	Fabaceae	'Arrow Poison'
02.	*Breynia oblongifolia*	Phyllanthaceae	'Snow Bush', 'Coffee Bush'
03.	*Brya ebenus*	Fabaceae	'Cocus Wood'
04.	*Calliandra tweedi*	Fabaceae	'Powder Puff'
05.	*C. brevipes*	Fabaceae	'Powder Puff'
06	*Duranta plumeri*	Verbanaceae	'Blue Duranta'
07.	*Ixora coccinea*	Rubiaceae	'Ixora'
08.	*Indigofera tinctoferia*	Fabaceae	'Indigo, 'Neel'
09.	*Galpinia transvaalica*	Lythraceae	'Galpinia', 'Wild Pride of India'
10.	*Gmellina hystrix*	Verbanaceae	'Parrot's Beak'
11.	*G. eliptica*	Verbanaceae	'Gmellina'
12.	*Jatropha podagrica*	Euphorbiaceae	'Gout Plant', 'Buddha Belly Plant'
13.	*Jasminum nudiflorum*	Oleaceae	'Winter Jasmine'

DIFFERENT STYLES OF BONSAI

Slanting (Zig-Zag)

Group Planting

Group Planting on Rocks

Informal Upright (Curved Trunk)

Informal Semi-Upright

Informal Upright (*Ficus* sp.)

Slanting Style (*Ficus* sp.)

Informal Upright (*Casuarina* sp.)

Formal Upright (Base Root Modification)

Formal Upright (*Ficus* sp.) with Crumbled Roots

Formal Style (Stem Crisscrossed)

Informal Upright

Formal Upright (Multiple Heads)

Informal Upright (*Ficus* sp.)

Semi-Cascade Style

Informal Slanting Style with Bridge

Informal Style (Ficus 'Long Island')

Formal Upright (*Adansonia* sp.)

Group Planting (*Ficus benjamina*)

Formal Upright (*Tamarindus* sp.)

Formal Upright (*Adansonia* sp.)

Slanting Style (Multiple Trunk)

Slanting Style (Single Stem)

Formal Upright Style

Semi-Cascade Style

Wind Swept Style

Formal Upright

Informal Upright

Style - Group Planting

Semi-Cascade Style

Slanting Style

Informal Upright (Leaning Left)

Informal Upright

Informal Upright (Leaning Right)

Semi-Cascade Style

Recurving Style

Formal Upright Style (Multiple Trunk)

Sl.No.	Botanical Name	Family	Common Name
14.	*Jacquina racemosa*	Primulaceae	'Bracelet Wood', 'Torch Wood'
15.	*Malpighia coccigera*	Malphigiaceae	*Malpighia*
16.	*Milletia ovalifolia*	Fabaceae	'Jewls on String'
17.	*Miscanthus sinensis var. Zebrinus*	Poaceae	'Chinese Silver Grass'
18.	*Malus domestica* 'Red Jade'	Rosaceae	'Apple'
19.	*Murraya exotica*	Rutaceae	'Kamini', 'Orange Jesmine'
20.	*Memecylon edule*	Malastomaceae	'Kaayam', Delek Bangas'
21.	*Olea europea*	Oleaceae	'Europian Olive'
22.	*O. africana*	Oleaceae	'African Olive', 'Indian Olive'
23.	*Petria arboria*	Verbanaceae	'Purple Wreath'
24.	*Phylanthus nivosus*	Phyllanthaceae	'Snow Bush'
25.	*Premna microphylla*	Lamiaceae	'Japanese Musk Apple',Japanese Neem Tree'
26.	*Psidium guajava*	Myrtaceae	'Guava'
27.	*Punica granatum*	Lythraceae	'Anar, 'Pomegranate'
28.	*Schefflera arbicola*	Araliaceae	'Umbrella Tree'
29.	*Serissa foetida*	Rubiaceae	'Snowerose', 'Japanese Boxthorn'
30.	*Rondeletia speciosa*	Rubiaceae	Rondeletia

Woody Creepers

Sl.No.	Botanical Name	Family	Common Name
01.	*Banisteria laurifolia*	Malpighiaceae	'Banisteria'
02.	*Bougainvillea* spp./cvs	Nyctagenaceae	'Bougainvillea', 'Paper Flower'
03.	*Combretum densiflorum*	Combretaceae	'Scarlet Combretum', 'Flame Vine'
04.	*Derris scandens*	Fabaceae	'Derris', 'Jewel Vine'
05.	*Hiptage madablata*	Malpighiaceae	'Rangoon Creeper', 'Chinese Honey Suckle'
06	*Petrea volubilis*	Verbanaceae	'Queen's Wreath', 'Purple Wreath'
07.	*Rupallia grata*	Apocynaceae	Climbing Olender', 'Indian Rubber Vine'

Sl.No.	Botanical Name	Family	Common Name
08.	*Stigmaphylon ciliatum*	Malpighiaceae	'Orchid Vine', 'Brazilian Golden Vine'
09.	*Tristellateia australasiae*	Malpighiaceae	'Australian Gold Vine'
10.	*Wisteria sinensis*	Fabaceae	'Chinese Wisteria'

9

PLANTING, PRUNING AND TRAINING

Planting

After selection of plant and container, planting should be done. As a first step, the container needs to be checked for presence of proper drainage hole. Potting mixture according to the plant species selected, need to be prepared for use. Planting of bonsai in trays or any other containers is similar to transplanting of other plants. Depending upon the style, the number of plants and shape of the container, place of planting (centre/ one third length of the tray/corner) is to be decided.

The potting mixture is prepared by mixing garden soil, leaf mould in 3:1 ratio besides *neem cake* and bone meal for providing slow releasing nutrients. The mixture should have good water holding capacity and drainage besides a balance nutrient status. The mixture should have sufficient nutritional strength to sustain the shock of repeated pruning, pinching, clipping *etc.* The mixture may have other ingredients plant nutrients as per plant species.

Planting in trays is done as per style selected which decides number of plants and its position in the tray. It may a single or multiple plants in a tray. The best time of planting is monsoon for evergreen plants or February-March for others. However, plantation can be done during moderate period also. After planting, irrigate gently and thoroughly. Extra water should be drained out through the holes. The containers/trays are then kept in semi–shady place for 7-10 days for root setting and, subsequently, shifted to lighted place gradually for proper growth.

Pruning

After a month or so, when vegetative growth has taken place, light pruning, pinching and training is started in order to shape the plants as per selected style. It takes considerable time to take proper shape as the growth rate remain restricted in shallow containers. The purpose of pruning is to provide desired shape to fulfill the style adopted. Frequency of pruning and type, light or heavy, depend on the growth rate, type of plant, season *etc.* Pruning encourages lateral branching in a desired direction. Younger plants need more frequent pruning than established older specimens.

Training

This is mainly done by using wires and the branches are curved and angled into desired direction, as per style adopted. Training should be done when the plants are in young stage, as the molding of the branches is easier at that stage. The entire process of bonsai making involving pinching, pruning, wiring and tying is a continuous repetitive process. One should have artistic perception, patient and skill to do the job. Creation of a perfect Bonsai specimen may take several years. The technique should be adopted at younger stage of growth when the trunk and branches are flexible and molded to desired shape. Copper wire of different gauge is used according to the purpose. Fixation of wire and tying is a technical job and to be done in a skilled manner, as per style followed.

Combination of pruning, training, wiring, pinching culminate into an artistic shape of bonsai.

10

CARE AND MAINTENANCE

This is most important how one cares and maintains the Bonsai specimens round the year. Proper procedure and a schedule of maintenance work, help to maintain the plants in a better way. Relatively new plants in Bonsai form require special treatments initially for few years until these take desired shapes. On the other hand, well grown and finished Bonsai require routine up-keeping by irrigation, pinching *etc.*

Bonsai needs intensive care and proper maintenance for keeping the plants in shape and attractive form. Pinching, pruning and tying are most commonly practiced techniques and frequently undertaken.

Small trees grown in containers, like bonsai, require specialized care. Unlike house plants and other subjects of container gardening, tree species in the wild, in general, grow roots up to several meters long and root structures encompassing several thousand liters of soil. In contrast, a typical bonsai container is less than 25 centimeters in its largest dimension and 2-10 liters in volume. Branch and leaf (or needle) growth in trees is also of a larger scale in nature. Wild trees typically grow tall when mature, whereas the largest bonsai rarely exceed one meter. These size differences affect maturation, transpiration, nutrition, pest resistance, and many other aspects of tree biology. Maintaining the long-term health of a tree in a container requires some specialized care techniques.

Repotting

Well established old plants need to be repotted by removing excess roots and replacing old potting mixture with fresh one. The fresh potting mixture is prepared by mixing soil, leaf mould, neem cake, bone meal *etc.* so that it release plant nutrient slowly in stead of quick vegetative growth. Repotting accelerates new growth, better flowering and fruiting. Consequently, the specimens become more appealing. Best period for repotting is March-April for deciduous type and July-August for evergreen type. It should be done at regular intervals depending upon the tree species, age and necessity.

The number of plants, its position and decoration should remain as it was before repotting. The plants should be kept under partial shade after repotting for a week or 10 days depending upon the species. Irrigation after repotting should be in the form of misting, no high pressure water sprays.

Watering

Must be regular and relate to the bonsai species requirement for dry, moist, or wet soil. It can sometimes be required several times per day. Soil condition and weather should be considered before deciding irrigation schedule. Sufficient amount should be added while irrigating so that extra water is drained out through the whole for better aeration. In summer season frequency of irrigation and amount of water added are more in comparison to winter and rainy season. Sometimes, spraying of water on the leaves helps cleaning of the foliage, removal of dust and light irrigation also.

Weeding

Sometimes weeds and unwanted vegetation grow in the tray after repeated irrigation. These should be removed time to time.

Plant Protection

Diseases and pests attack plants in different season and cause damage to the plants, especially foliage. This causes leaf damage and looks ugly. Therefore, spraying of insecticides and fungicides should be conducted at regular intervals to keep the Bonsai specimens free of diseases and pests.

Location and Over-wintering

When the bonsai is kept outdoors, different species require different light conditions. Few of the traditional bonsai species can survive inside a typical house, due to the usually dry and warm indoor climate. Therefore, display location is important and to be considered. Deciduous plants should be placed in warmer and lighted place during winter.

Leaf Trimming/Defoliation

Generally it is practiced for tree species having large leaves for reducing leaf size after regeneration. This selective removal of leaves is for the varieties of deciduous tree.

Pruning

Removal of undesirable branches, twigs to retain the desired shape. This is repeated several times during the growth period and whole life of Bonsai. Frequency and intensity of pruning depend on plant species selected, nature of growth, season *etc.*

Clamping

Using mechanical devices for shaping trunks and branches.

Root Pruning

This is a routine operation in bonsai making, done when repotting is carried out. Over grown, crumbled roots of thinner diameter are removed by pruning keeping intact the main tap roots. This encourages new root formation and fresh vegetative growth and flowering.

Dos and Don'ts

- Do not add heavy nitrogenous fertilizers which accelerate quick vegetative growth requiring repeated pruning and pinching.
- Well grown specimens, that have already taken shape, should not be pruned heavy.
- No foliar application of NPK which accelerate quick vegetative growth.
- Before pruning and pinching consider the type of plant, flowering season, active growth period for best display.

- Take proper protection against pests and diseases by regular spraying of insecticides/fungicides.
- Do not over irrigate the plant which causes yellowing of leaves and fungal diseases.
- Protection of plants from scorching sun as per light requirement.

Seasonal Operations

Maintenance of Bonsai round the year is a specialized job and need knowledge on growing different types of plants, their climatic requirement, cultural practices to be done. Growth pattern of the trees and plants in different season differs and accordingly the requirement of pruning, training, watering *etc.*

Young plants need special attention while older specimens should be treated separately. As the climatic pattern (temperature, light, and humidity) varies in different season, the cultural operation also varies. Therefore, the maintenance job has been categories into six different following seasonal operations.

Summer Season (April-June)

- Temperature remains high in sub-tropical and tropical climate together with low humidity.
- Plants remain in active growth phase.
- Frequency of irrigation will be more together with pruning, pinching *etc.*

Rainy Season (July-August)

- Temperature moderate.
- High rainfall.
- High humidity.
- Frequency of irrigation will be less but pruning, pinching requirement will be more.
- Right season for repotting, deep pruning.

Autumn Season (September-October)

- Temperature moderate.

- Less rainfall.
- Moderate humidity.
- Frequency of irrigation will be less.
- Right season for repotting, deep pruning.

Late Autumn (November)

- Temperature gradually drops.
- Less rainfall.
- Less humidity.
- Frequency of irrigation will be less.
- Not right season for repotting, deep pruning.
- Regular operations like watering, plant protection to continue.

Winter Season (December-January)

- Temperature remains low.
- Less rainfall.
- Less humidity.
- Frequency of irrigation will be less.
- Not right season for repotting, deep pruning.
- Plants remain in inactive growth phase.
- Regular maintenance to continue.

Spring Season (February-March)

- Temperature starts rising.
- Plants start active growth after dormancy.
- Regular maintenance to continue.
- Right season for repotting, training *etc.*

11

TOOLS AND IMPLEMENTS

Making a perfect bonsai involves several technical operations and cultural practices. For technical operations a, set of tools and implements are required for doing all those operations in a perfect way. Selection of proper type of equipments is important for accurate execution of the cuts, bends, clippings, pruning *etc.*

For beginners, it is recommended that few basic tools may be acquired and used; like concave cutter, shear *etc.* For intensive and fine work special tools will be needed. Many tools and implements are available in market, for carrying out various operations. One has to select the purposeful one. Tools of Japanese make are of high quality but little costly. However, tools are suitable for long time use.

There is wide variety of tools and implements as follows:

1. Large Concave Cutter with straight blade	13. Thin Pruning Shaw
2. Small Concave Cutter with straight blade	14. Knives
3. Knob Cutter	15. Tweezer Spatula
4. Pliers	16. Grafting Knife
5. Pruning Shears	17. Root Hook
6. Standard Shear	18. Small Root Rake
7. Long Slim Twig Shear	19. Large Root Rake
8. Large Foldable Saw	20. Sickle Saw

9. Sickle Knife	21. Large Wire Cutter Small
10. Soil Scoops	22. Root Plier
11. Strong Standard Shear for Root Pruning	23. Wire bending Plier
12. Wire bending Plier	24. Wire Cutter

Miscellaneous Other Items Required

Raffia	Fusing Rubber tape and Gauze	Infusion Hose
Branch/Trunk Bending Lever	Screw Clamp	Rust Eraser and grind stone
Disinfectant	Gun oil	Camellia Oil
Nylon Brush	Brass Brush	Steel Brush
Tooth Brush	Coco brush	Bicycle Tube

Main Tools and Implements

Use of Tools and Implements

- **Saw, Cutter and Knife** – For cutting branches, trunk, roots which are thick or hard.
- **Tools for Repotting or working on Roots** - Sickle Knife, Sickle Saws, Shears.
- **Wiring** - Wire bending Plier, Wire Cutter.
- **Removal of Deadwood** – Branch splitter (Sharp Plier), Slim Chisel, Knives

Each and every tools, implements and other items has specific purpose and to be used accordingly.

Miscellaneous Items (Tools and Implements)

12

DISPLAY

The most important thing is how to display bonsai specimens for exploiting their beauties as indoor plants. There are several ways to display and to beautify interiors with these lovely specimes. Before finalizing display, one should consider the following points which will help for an effective display.

- Inspection of the interiors in order to find out best place for display.
- Availability of light and position of the objects (furniture, show case, stairs, steps, entry, exit *etc.*
- Size of the specimen and finding a befitting place of display.
- Selection of the specimens as per place of display (flowering specimens need lighted place).
- Placing the specimens at prominent place so that visitors' attention is drawn.
- Use of appropriate pedestal for display.
- Some suggestive places – both sides of the main entrance of the interiors, staircase landing, lobby, drawing room, on tea/dining table (small specimens).

Specimens kept at one place should be shifted/rotated from one place to another, less lighted area to bright area. This helps the specimens to maintain proper form. Otherwise, too long keeping at one place may lead to de-shaping, yellowing of leaves and loosing luster.

Display in Indoor

Display of Bonsai plants in indoor is a specialized job and one has to consider many things for appropriate display. A bonsai display presents one or more bonsai specimens in a way that allows a viewer to see all the important features of the bonsai from the most advantageous position. That position emphasizes the front view, which is most attractive part of the Bonsai. It places the bonsai at a height that allows the viewer to imagine the bonsai as a full-size tree seen from a distance. The placement should not be neither too low so that it does not escape the attention of the viewers. It should be shown as most focused way, as if in art gallery. The right height, isolation, plain background and should be devoid any other things which distracts attention. In indoor display, Bonsai plants are kept temporary, lasting a week or so, as most bonsai are intolerant of indoor conditions for long and lose vigor rapidly within the house. Therefore, shifting the plants one place to another and changing of plants at regular interval is advised.

Display at Outdoor

While placing at the outdoor for display, aesthetic point should be considered first besides other nearby structures and objects. Many outdoor displays are semi-permanent, with the bonsai trees in place for weeks or months at a time. In order to avoid any damage of the trees, selection of right place considering sunlight availability is must followed by good maintenance practices. Most common way is to display on the bench or individual pedestal. The arrangement may be formal in line or informal best suited the display area.

Where space allows, outdoor bonsai specimens are spaced far enough apart that the viewer can concentrate on one at a time. When the trees are too close to each other, aesthetic discord between adjacent trees of different sizes or styles can confuse the viewer.

Display in Exhibition

Display of Bonsai in exhibitions is another way of exhibiting the specimens in different patter. If there is any theme, the specimens should be grouped as per theme *viz.*, foliage, fruit bearing, flowering types *etc.*

Exhibition displays allow many bonsai to be displayed in a temporary exhibition format, typically indoors, as would be seen in a bonsai design competition. To allow many trees to be located close together, exhibition displays often use a sequence of arrangement.

Display in small alcoves, each containing one pot, is a standard method. The walls or dividers between the alcoves make it easier to view only one bonsai at a time. The back of the alcove is a neutral color and pattern to avoid distracting the view. The bonsai pot is almost always placed on a formal stand considering its size, design and synchronization with other components.

13

COMMERCIAL IMPORTANCE

Commercial importance of Bonsai in horticultural and nursery trade is enormous. The most common plant of Bonsai is *Ficus* species. The cost of a nursery plant of *Ficus* or any other plant for Bonsai is INR 100.00 (less than a Dollar) or less. However, the same plant after value addition and trained as Bonsai fetch price much more than the nursery plant. But preparing Bonsai from a nursery plant requires time, know-how, maintenance at least 2-3 years for making a basic specimen.

What to do for Commercial Importance

The approach for making Bonsai for commercial purpose is different, if it is grown by an individual or a nurseryman. However, the basic requirements are same and depend on the following point for its commercialization.

1. **Selection of Right Type of Plant Species** - Select those which are easy to convert as Bonsai. In stead of selecting a very small plant, select a grown up one which is suitable for Bonsai making. This saves time for the transformation and gives quick result. Naturally dwarf plant species should be selected *e.g. Ficus* 'Long Island'. Other species for quick result are – *Acacia* spp., *Ficus* spp., *Prosopis cineraria, Psidium guajava* (Chinese Guava), *Punica granatum, Carissa carandus, Citrus microcarpa etc.*

2. **Selection of Right Type of Containers** – Bonsai for commercial purpose should have a unique and attractive container which is decorative to fetch high price.
3. **Style**: This should be unique according to the plant species selected.
4. **Landscape/Base Decoration** – In addition to the plant species, the base decoration is important especially when large pot is used.
5. **Height** – Should have medium height and easily recognizable from distance.
6. **Over All Appearance** – This is most important criteria for fetching high price. The specimen should be perfect in all respect – style, training, decoration *etc.*

The pricing of the Bonsai specimen depend on its overall appearance, attractiveness, boldness, foliage beauty, age expression through prop roots, branches *etc.* The price range may be several thousand rupees to lakhs depending upon the specimen and its look. Therefore, making Bonsai is a profitable trade in nursery business in comparison to nursery for common plants.

14

BONSAI SOCIETIES AROUND THE WORLD

Originated in China, excelled in Japan, the bonsai technique – the art of growing trees and plants in dwarf form in an aesthetic way is now spread over the world. Its ornamental looks, aesthetic appearance keep onlooker mesmerizing. Bonsai is now a living artifact, a show piece to display in every garden, indoor and outdoor. This tremendous popularity has been achieved due to hard and sincere efforts of many renowned individuals and societies who effortlessly carried over the bonsai flag since last so many decades. Amateurs, hobbyists, intellectuals, specialists join hands for the promotion of bonsai in many forms *viz.*, shows, exhibitions, training, workshops under different banners in separate countries. But the objective was popularization of Bonsai in large scale.

Following societies in different countries were instrumental for promoting and popularizing bonsai culture.

Australia

- Bonsai in Australia, Muni Ponds.
- The Bonsai Society of Australia Inc.
- Bonsai Society of Victoria.

- The Bonsai Society of Sydney.
- South Australian Bonsai Society.

China

- National Bonsai Association, Beijing.
- Sanghii Bonsai Association, Shanghai.
- Guan Zhao Bonsai Association, Guan Zhao.

India

- The Agri-Horticultural Society of India, 1, Alipore Road, Kolkata, 700027, West Bengal.
- Bonsai Study Group of the Indo-Japanese Association, Worikon House, Mumbai, Maharashtra.
- Indo-Japanese Bonsai Society, Dr. Annee Basant Road, Worli, Mumbai, Maharashtra.
- Indian Bonsai Association, Delhi.
- Avadh Bonsai Association, Lucknow, Uttar Pradesh.

Japan

- Nippon Bonsai Association, Tokyo.
- Kawaguchi Bonsai Association, Kagawa.

United Kingdom

- British Bonsai Association, 23, Nimrod Road, Streatham, London, SW 16.
- British Bonsai Society, 35, Clevedon Road, Failand, British, Avon.
- Belfordshire Bonsai Society, 65, Richmond Hill, Beds LU 27 JG.
- Croydon Bonsai Club, 7, Mulbery Lane, Crydon, Surrey CRO 6 QR.
- Manchester Bonsai Society, 160 Chedlei Old Road, Stockport, Cheshire.
- Yorkshire Bonsai Association. 187 West Park Drive West, LS and 2E.
- Sussex Bonsai Society, 98, Fair Field Road, Sussex RH 158NT.

- National Bonsai Society, 24, Wiston Road, Liverpool, L47XJ.
- Middelsex Bonsai Society, 54, Tregenna Avenue, Hrro, MDH.A 28, QS.
- Federation of British Bonsai Society, London.

United States of America

- Bonsai Institute of California, Los Angels.
- Bonsai Society of Greater New York.
- Mid Atlantic Bonsai Society, New York.

15

IMPORTANT LITERATURE

1. Adams, Peter D.1981. The Art of Bonsai. Ward Lock Ltd. pp. 134.
2. Anonymous. 1989. Bonsai of the World. World Bonsai Friendship Federation, Japan, pp. 96.
3. Anonymous. 1989. Classic Bonsai of Japan, Nippon Bonsai Association, Kodansha International, pp.188.
4. Bonsai Empire and Walter Pall and Mauro Stemberger. 2014. Bonsai: A Beginners Guide, pp. 101.
5. Bose, T.K. and Bhattacharjee, S.K. 1980. Garden Plants, Oxford and IBH Publishing Co., Calcutta, New Delhi, Bombay, pp.273.
6. Bose, T.K., Maiti, R.G., Dhua, R.S., Das, P. 1999. Floriculture and Landscaping, Naya Prokash, Calcutta,India, pp.757.
7. Chan, Peter.1987. *Bonsai Masterclass*. Sterling Publishing Co., Inc., New York, USA.
8. Chidamian, C. 1960. Bonsai Miniature Trees, D. Van Nostrand Company, New Jersey, Toronto, London, New York, pp. 96.
9. Dhanda, Leela. 1996. Bonsai Culture, Oxford and IBH Publishing, New Delhi, India, pp.142.
10. Gustafson, Herbert L. 1995. Miniature Bonsai. Sterling Publishing Company, Inc., New York, USA, pp. 119.

11. Jana, B.L. 2018. Bonsai – The Living Art. *Mehanati Prakashan*, Hooghly, West Bengal, India.
12. Koreshoff, D.R. 2010. Bonsai: Its Art, Science, History and Philosophy, Boolarong Publications, Australia, pp.255.
13. Lesniewicz, Paul.1996. *Bonsai in Your Home*. Sterling Publishing Company, USA, pp.107.
14. Lewis, Colin.2001. *The Art of Bonsai Design*. Sterling Publishing Company, Inc.,: New York, USA, pp. 44–51.
15. Mishra, K.K. 2019. Floriculture : Principles and Practices, Brillion Publishing, New Delhi, India, pp. 327.
16. Morris, P.L. and Saphire, S.W. 2008. Growing Bonsai Indoors, Brooklyn Botanic Garden Allb-bRegion Guide, U.S.A, pp. 120.
17. Naka, J.Y. 1998. Bonsai Techniques II, Bland Ford Press, California, U.S.A.
18. Naka, John Yoshio. 1973. *Bonsai Techniques I*. Bonsai Institute of California. pp. 123–124.
19. Nippon Bonsai Association. Classic Bonsai of Japan. p. 144.
20. Norman, Ken.2005. Growing Bonsai: A Practical Encyclopedia. Lorenz Books. pp. 176–177.
21. Parekh, J. and Parekh, N. 1979. Tropical Bonsai, Group of the Indo-Japanese Association, Bombay, India.
22. Parekh, J. and Parekh, N. 2009. Wonder World of Bonsai, Vakils Publisher, Bombay, India.
23. Paul, L. 1994. Bonsai in Your Home. Sterling Publishing Co, New York, U.S.A.
24. Pessey, C. and Samson, R. 1993. Bonsai Basics: A Step-By-Step Guide to Growing, Training and General Care, Sterling Pub. Co. Inc, New York, USA.pp.17.
25. Prescott, David, 2017. The Bonsai Handbook, pp.421
26. Roy, R.K.2013. Bonsai *In*: Fundamentals of Garden Designing, New India Publishing Agency, New Delhi, pp.471-488.
27. Roy, R.K. 2020. Bonsai *In*: Home Gardening – A Guide Book, Roy's Green and Gardens Foundation, pp.165-170.

28. Valsala Kumari P.K., Rajeevan, P.K., Sudhaadevi, P.K. and Geetha, C.K. 2008. Flowering Trees, New India Publishing Agency, New Delhi, India, pp. 288.
29. Yamada, Tomio.2005. Fundamentals of Wiring Bonsai. *International Bonsai* (4): 10–11.
30. Yoshimura, Yuji. 1991. Modern Bonsai, Development of the Art of Bonsai from an Historical Perspective, Part 2. *International Bonsai* (4): 37.
31. Yuji Yoshimura and Barbara M. Halford.1957. *The Art of Bonsai: Creation, Care and Enjoyment.* Tuttle Publishing, North Clarendon VT USA. pp. 65–66.

16

SOURCE OF PLANTS/BONSAI

In India

Andhra Pradesh

1. Classic Farms and Nurseries, Regd. Office: G-3, Ashraya Apartments, Mettuguda, Secunderabad – 500 017, Andhra Pradesh.
2. Famous Flower Centre, S. No. 15, New Osmangunj, Hyderabad – 500 012, Andhra Pradesh.
3. Madhu Babu Flowers, 285/B/30, Old CIB Quarters, Khairatabad, Hyderabad 500 004, Andhra Pradesh.
4. Madhu Farms, H.No. 16, New Balaji Nagar, Near Karmanghat Hanuman Temple, Karmanghat, Hyderabad- 500 079, Andhra Pradesh.
5. S.S. Flower, 1-10-27/2, Lane Opp. Ashok Nagar, Park Street No. 3, Hyderabad – 500 020, Andhra Pradesh.
6. Sree Rajkumar Nursery and Gardens, 102, Whitefield, Kondapur, Hydrabad, Andhra Pradesh.
7. Sri Ram Nursery, Behind Rahul Automobiles, Begumpet, Hyderabad – 500 016, Andhra Pradesh.

8. Sri Satyanarayana Nursery, Kadiyapulanka, Rajahmundry, E.G. Dist., Andhra Pradesh - 533 126.
9. Sri Veera Hanuman Nursery, Kadiapulanka, Kadiam Mandal E. G. DT, Andhra Pradesh - 533 126.
10. Sri Vijaya Durga Nursery, Kadiyapulanka – 533 126, Rajmahmundry, E.G. Dist., Andhra Pradesh.

Delhi

1. Argosy, 97, Yashwant Place, Chanakyapuri, New Delhi – 110 021.
2. Capital Nursery, Badarpur, Near Sai Mandir, E Block, Hari Nagar Part 2, Delhi - 110044.
3. Ever Green Nursery, Karbala, Jor Bagh Road, New Delhi – 110 003.
4. Garden Blossom, 9, C.S.C. Market, Kaka Nagar, Dr. Zakir Hussain Road, New Delhi – 110 003.
5. Govt. Sunder Nursery, Nizamuddin, New Delhi - 110 013.
6. Greenways Nursery, Office: 15, Aurobindo Apartments, Adhchini, New Delhi – 110 074.
7. Griha Vatika, D-I/1159 Vasant Kunj, New delhi, New Delhi-110070.
8. Masjid Nursery, Pandara Road, New Delhi – 110 092.
9. Rajdeep Agri Products Pvt. Ltd., 3279/1, Ranjeet Nagar, New Delhi – 110 008.
10. Rajdhani Nursery, Karbala, Jor Bagh Road, New Delhi-110003.

Gujarat

1. Guj. Horti. Culture, New Law College, Panchvati Road, Ellisbridge, Ahmedabad-380007, Gujarat.
2. Patel Farm and Nursery, Udvada (R.S.), Valsad, Gujarat.
3. Sagar Nursery, PN Marg, Bedi Bandar Road, Near Ashutosh Petroleum, 11, Jamnagar-361002, Gujarat.
4. Shri Ramesh Nursery, Pirana Road, Beharampur, Ahmedabad-380022, Gujarat.
5. Surendra Nursery, Navayard Chhani Road, Fatehgunj, Vadodara-390002, Gujarat.

Himachal Pradesh

1. Dhauladhar Flower Nursery, V.P.O 53 miles, Kangra-176056, Himachal Pradesh.
2. Green World Nursery and Landscaping, National High Way 21A, Bhud, Solan, Baddi-174101, Himachal Pradesh.
3. Lajwanti Nursery, 124, Vir Bhawan, Shyam Nagar, Dharamsala-176215, Himachal Pradesh.
4. Shivalik Agro Engineers, Near DAV School, Kather Bypass Road, Solan – 173 213, Himachal Pradesh.
5. V.K. Flower Nursery, Dharamsala Road, Gaggal, Kangra-176209, Himachal Pradesh.

Karnataka

1. Balaji Nursery and Garden, Lalbagh, Siddapura, Bangalore – 560 011, Karnataka.
2. Indo-American Hybrid Seeds (I) Pvt. Ltd., 17th Cross, K.R. Road, Banashankari II Stage, Bangalore – 560 070, Karanataka.
3. Krishnendra Nursery, 9-159, Lalbagh Siddapura, Jayanagar 1st Block, Bangalore – 560 011, Karnataka.
4. KSG and Sons, 177, 5thMain road, Chamrajpet, Bangalore – 560 018, Karnataka.
5. The Nurserymen Co-operative Society Ltd., Lalbagh, Bangalore – 560 004, Karnataka.

Maharashtra

1. Gardenia Plants Nursery, Plot No. 90, Floriculture Park, MIDC, Talegaon, Pune, Maharashtra.
2. Khushboo, 206, Nana Path, Pune – 411 002, Maharashtra.
3. Krishi Kranti Kendra, Shani Mandir Road, Sitabuldi, Nagpur, Maharashtra.
4. Namdeo Umaji Agritech (I) Pvt. Ltd., 1205/4, Alankar, 1st Floor, Opposite Sambhajii Park, Shivaji Nagar, Pune – 4, Maharashtra.
5. Oasis Floral Products, 605, OM Chambers, Kemps Corner, Mumbai – 400 036, Maharashtra.

6. Poocha Seeds Pvt. Ltd., Near Sholapur bazaar, Pune – 411 001, Maharashtra.
7. Rose Nursery, Siddhivanayak, Krushnakutir Buldg., Shop No. 2, Tapodham Colony, Talegaon Station, Dist Pune, Maharashtra.
8. Srividya Nursery, Bori Aindi (Urulikanchan), Near I.T.I. College, Boribhadak, Bori Aindi Road, Maharashtra.
9. Tukai Exotics, Gat. 253, Opp. H.P. Terminus, Pune-Solapur Road, Loni Kalbhor, Pune-412201, Maharastra.

Punjab

1. Bauscape Farms, J-26, Sarabha Nagar, Ludhiana-141 001, Punjab.
2. Flora Landscape and Nursery, Near I.B.P Petrol Pump, Barnala Road, Sangrur -148001, Punjab.
3. New Janta Nursery, Ludhiana Road, Malerkotla- 148023, District - Sangrur, Punjab.
4. Plantsman's Seeds, Rajbaha Road, Patiala – 147 001, Punjab.
5. Prakriti Landscapes, 4020, Sector 46-D, Chandigarh – 160 047, Punjab.

Rajastan

1. Hariyali Kunj Nursery, Plot No. 24-25, Shanker Nagar, Opp. Road No-1, V.K.I Area, Siker Road, Jaipur-302013, Rajastan.
2. M.D. Nursery, Naib Ji Ka Bagh, Moti Doongri Road, Opp. Heera Panna Complex, Jaipur-302 004, Rajastan.
3. Shanghralaya Art, Girish Nagar, Agra By-pass, Jaipur, Rajasthan.
4. Shri Shyam Nursery, D - 618, Murlidhar Vyas Nagar, Bikaner-334001, Rajastan.
5. Vissu Nursery, Shahid Bhagat Singh Colony, Banswara-327001, Rajastan.

Uttrakhand

1. Dev Bhumi Nursery, 61, Suman Nagar, Dharampur, Dehradun, Uttarakhand -248001.

2. Hortico Nursery, Floridale School Campus, Kargi Chowk, Dehradun. Uttrakhand -248001.
3. Indo-Dutch Horticulture Technology, Chaffi, Bhimtal, Uttrakhand.
4. Luxmi Nursery and Seeds Store, Panditwari, P.O. Premnagar, Dehradun-248007, Uttarakhand.
5. Pratap Nursery and Seed Stores, Prem nagar, Dehradun, Uttrakhand.

Uttar Pradesh

1. Friends Rosery, B-110, Mahanagar, Lucknow, Uttar Pradesh.
2. Garden Centre, Shivalik Enterptises, Nadar Ganj, Lucknow, Uttar Pradesh.
3. Phulwari, Vidhan Sabha Marg, Lucknow/A-74, Sector 27, Noida, Uttar Pradesh.
4. Pushpam Nursery, E-18, Liberty Colony, Sarvoday Nagar, Lucknow, Uttar Pradesh.
5. Sandeep Nursery, B-69, Mandir Marg, Mahanagar, Lucknow, Uttar Pradesh.
6. Vatika, C-986/B, Mahanagar, Lucknow 226 006, Uttar Pradesh.
7. Saurabh Nursery, Lucknow, Uttar Pradesh.

West Bengal

1. Arambagh Nursery, 59 B, Chowringhee Road, Kolkata – 700 020, West Bengal.
2. DBN Agro-Horticulture (P) Ltd., Regd. Office: Vill – Sripur, P.O.- Sripur Bazar, Dist – Hooghly – 712 514, West Bengal.
3. Kamal Nursery, Andul-Mouri, Howrah, West Bengal.
4. Sahayog Hortica Pvt. Ltd., Bakhrahat Road, Samukpota, P.O. Kanganberia, District 24 parganas (S), West Bengal - 743503.
5. The Agri-Horticultural Society of India, 1, Alipore Road, Kolkata – 200 027, West Bengal.
6. The Deb Narayan Nursery, Sukhdevpur, Bishnupur – 743 503, South 24 Parganas, West Bengal.

7. The Mondal Nursery, Bakhrahat, 24 PGS (S), West Bengal.
8. The Rainbow Garden, Mahamayapur, 24 PGS (S), West Bengal.
9. Horticulture Development Centre, Judges Court Road, Kolkata-700 027, West Bengal.

Availability of Bonsai in Other Major Countries

Australia	Singapore
China	Taiwan
Korea	Thailand
Honkong	United Kingdom
Japan	United State of America
Indonesia	Vietnam
Malaysia	
Philippines	

INDEX

9 789359 193434

Printed at Repro India Ltd.